Real Science-4-Kids

Kogs-4-Kids

Chemistry Connects to

Language

Workbook Level I A

Rebecca W. Keller, Ph.D.

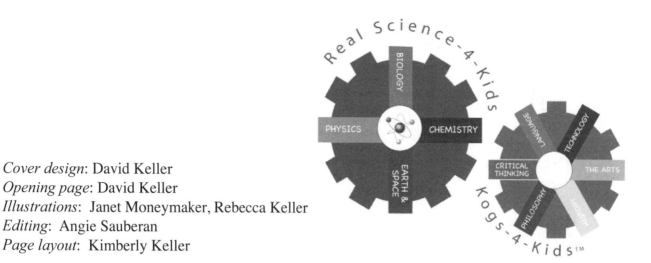

Cover design: David Keller
Opening page: David Keller
Illustrations: Janet Moneymaker, Rebecca Keller
Editing: Angie Sauberan
Page layout: Kimberly Keller

Real Science-4-Kids/ Kogs-4-Kids™ : Chemistry Connects to Language: Level I A

ISBN: 9780979945946

Published by Gravitas Publications, Inc.
P.O. Box 4790
Albuquerque, NM
87196-4790
www.gravitaspublications.com

Printed in the United States of America

Special thanks to G.E. McEwan for valuable input.

Gravitas
Publications Inc.

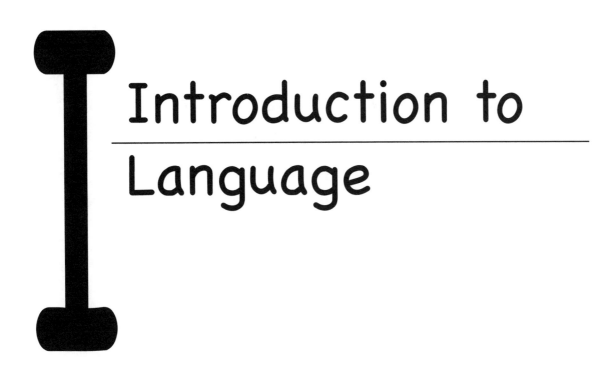

I Introduction to Language

I.1 The language of science

Have you ever noticed that scientists use all kinds of fancy words like *nucleosynthesis* [nü-klē-ō-sin´-thə-səs] and *photoelectric photometry* [fō-tō-i-lek´-trik fō-tä´-mə-trē]? Many of the words that scientists use are indeed long and difficult to pronounce. However, these words have been carefully selected by scientists that have put the field of science into verbal language. Each scientific word means a particular thing. There is a *language* to science.

I.2 Latin and Greek roots

If someone had to memorize all of the words that scientists use, it would be a difficult task. However, most of the words that are used in science come to us from two languages: Latin and Greek. Many of the words you encounter will have Latin or Greek *word roots*.

A word root is that part of a word that is derived from another word. For example, the word "biology" comes from the Latin word *bios*, which means "life," and *logy*, which means "study of." So biology means the "study of life". The word tree (on the previous page) illustrates several different words and their Latin or Greek word roots. You can see that the words on the branches are the word roots, and that those on the leaves come from these roots. In fact, many of the languages that people speak have Latin or Greek word roots. English, Spanish, Portuguese, German, and even Romanian all have some words that are similar because some words in each of these languages come from Latin or Greek word roots. For example, the English, Spanish, Portuguese, French, and Italian words for school all come from the Latin word *schola*.

"school"	language
schola	**Latin**
escuela	Spanish
escola	Portuguese
scuola	Italian
school	English
école	French

We can learn a lot about languages by learning Latin and Greek word roots.

I.3 How this book works

This workbook *connects* the discipline of science to the language of science. It will help you to understand the different subjects in science if you know about the language of science.

In the first section of each chapter, you will find the word root for a set of six English words.

For example, your word list may look something like this:

> deflate
>
> inflate
>
> flabellum
>
> flavor
>
> conflate
>
> afflatus

The word root can be two letters long, or three letters long, or even four or five letters long. You are to look for the two to five letters that are common in each word. This is the word **root**. For example, all of the words in this list have the following three letters in common:

> deflate
>
> inflate
>
> flabellum

(fla)vor

con(fla)te

af(fla)tus

We see that all of these words have the three letters "f," "l," and "a" in common, and that these letters make up the word root _f_ _l_ _a._

The exercises in the first section of each of the following chapters in this workbook are designed to get you thinking about the words on the list and the common word root. You are to try to *guess* the meaning of the root word and the meanings of the words on the list.

In the second section of each chapter, the meaning of the word root is defined for you. For example, we found that the word root for our example list was the three letter word root **fla**.

de**fla**te	**fla**vor
in**fla**te	con**fla**te
flabellum	af**fla**tus

We find out in this section that the meaning for the word root **fla** comes from the Latin word *flare,* which means "wind" or "to blow." We find out that all of the words have something to do with wind or blowing. The exercises in this section encourage you to try to define the words on your list *before* you look at the definitions. Guessing is good! It gets you thinking, even if your guesses are wrong.

The third section of each chapter defines all of the words. These definitions are taken from a number of different dictionaries, including Webster's Unabridged New Twentieth Century Dictionary 1972, the American Dictionary of the English Language 1828, and A Thesaurus of Word Roots of the English Language. Additional Latin or Greek prefixes are also given.

The meanings of the words in our example list are as follows:

deflate	1. to collapse by letting out air or gas. 2. to lessen the importance of, as with money. (*de* means *opposite*)
inflate	1. to blow full of air or to expand. 2. to raise the spirits of or to make proud. 3. to increase, as in raise beyond normal. (*in* means *in*)
flabellum	1. a large fan usually carried by the Pope. 2. in zoology or botany, the fan-shaped part or structure.
flavor	an odor, smell, or aroma *carried by the wind.*
conflate	1. to blow together, bring together, or collect. 2. to combine, melt, fuse, or join. (*con* means *together*)
afflatus	inspiration, a divine imparting of knowledge. (*af* means *to* or *toward*)

In the fourth section of each chapter, you are given the opportunity to match the definitions of the words to the list words. Here is an example of this matching exercise.

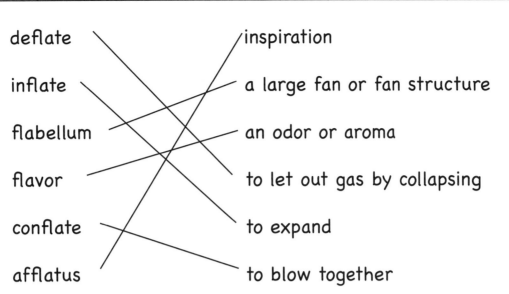

deflate — to let out gas by collapsing
inflate — to expand
flabellum — a large fan or fan structure
flavor — an odor or aroma
conflate — to blow together
afflatus — inspiration

In the fifth section of each chapter, you "test yourself." Now that you have learned the word root and the definitions of the different words on the list, you can see how well you remember them with a self test. Finally, in the last section of each chapter, you will be asked to write a story or several sentences using the words you have learned. Try to incorporate all of the words you have learned. Here is an example story.

While carrying a rather **deflated** balloon down the sidewalk, and his hand **conflated** with his mother's, little Johnny was losing his interest with his mother's conversation with Mr. Longs, who awkwardly carrying his **flabellum**, was nevertheless chatting whole-heartedly. The **flavor** of his chocolate sundae was wearing off and his patience was at an end. Afflating hard upon his mother, which was his favorite way of gaining her attention, his mother made her farewells and Mr. Longs, still awkwardly carrying his **flabellum**, moved on. Johnny was saved from the boring conversation at last. *--Written by Christopher Keller, age 8*

Now that you have learned how language connects to science, work through this workbook as you study Level I Chemistry, and most importantly..., have fun!

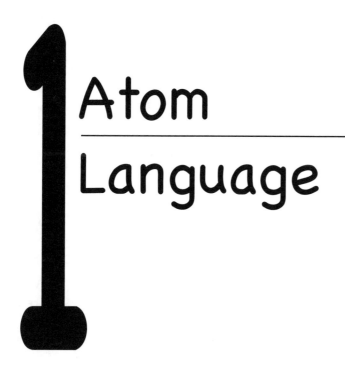

Atom

Language

1.1 Find the root

Look at the following words:

> atom
>
> anatomy
>
> diatom
>
> epitome
>
> entomology
>
> tome

There is a "cluster," or group of three letters, that is exactly the same in all six words. Can you find the cluster?

Circle the cluster that is the same in each word. Write the three letters that make up the cluster. _____ _____ _____

Now look at the words carefully. Because they all have a common cluster, they all have some similarities in their meanings.

Can you guess the meaning of the cluster?

_____

Can you write a definition for any of the words on the list?

_____

_____

_____

_____

1.2 Learn the root

Word Cluster

a**tom**   **tom**e

ana**tom**y   epi**tom**e

dia**tom**   en**tom**ology

All of the words above have a common word "root": **tom**. The word root **tom** comes from the Greek word *temnein*, which means "to cut." All of the words have something to do with the verb "to cut."

Now, knowing that the cluster **tom** comes from the Greek word *temnein,* try to guess the meanings of the words *before* looking at the definitions in the next section.

atom _____

_____

tome _____

_____

anatomy _____

_____

epitome _____

_____

diatom _____

_____

entomology _____

_____

1.3 Definitions

anatomy   the branch of science that deals with the structure
               of living things. Literally, the word anatomy means "to
               cut up." In science, sometimes animals are cut up to
               find out the position and structure of their parts. (*ana*
               means *up* or *again*)

atom   a small unit of matter that cannot be further cut. The
               Greeks thought that the atom was the smallest unit
               of matter, and that it could not be further divided.
               Today we know that atoms can be divided into smaller
               parts, but they are still called atoms. (*a* means *not*)

diatom   a small organism that has a wall that splits it into two
               symmetrical parts. Diatom literally means two atoms.
               (*di* means *two*)

entomology   the branch of biology that deals with insects. Insects
               have segmented bodies that are divided into parts.
               Hence, they look cut or divided. (*logy* means *study of*)

epitome   a short, brief statement or thing that represents
               the whole. For example, you might hear someone say
               "Michael Jordan is the *epitome* of basketball." (*epi*
               means *upon*)

tome   a large scholarly book. A tome is generally one volume
               that is part of a larger set.

1.4 Mix and match

Draw lines to connect the words with their meanings.

atom

anatomy

epitome

entomology

diatom

tome

a word or thing that represents the whole

a volume that is part of a larger set

a small creature that is separated into two parts by a wall

a small unit of matter that cannot be cut

the branch of science that deals with the structure of living things

the study of insects

1.5 Test yourself

Write the meanings next to the words below.

atom _____

_____

anatomy _____

_____

diatom _____

_____

epitome _____

_____

entomology _____

_____

tome _____

_____

Extra Words

Can you guess the meanings of the following words?

gastrotomy (*gastro* means *stomach*)

_____

xylotomy (*xylo* means *wood*)

_____

1.6 Using new words

Write sentences using each new word you have learned. If you wish, you may use the words to write a story.

_____

_____

_____

_____

_____

_____

_____

_____

_____

_____

_____

_____

_____

_____

2 Covalent Language

2.1 Find the root

Look at the following words:

> covalent
>
> valiant
>
> invalid
>
> equivalent
>
> valedictorian
>
> validate

There is a "cluster," or group of three letters, that is exactly the same in all six words. Can you find the cluster?

Circle the cluster that is the same in each word. Write the three letters that make up the cluster. _____ _____ _____

Now look at the words carefully. Because they all have a common cluster, they all have some similarities in their meanings.

Can you guess the meaning of the cluster?
_____

Can you write a definition for any of the words on the list?
_____

_____

_____

_____

2.2 Learn the root

Word Cluster

coval**ent**   **val**iant

in**val**id   equi**val**ent

valedictorian   **val**idate

All of the words above have a common word "root": **val**. The word root **val** comes from the Latin word *valere*, which means "strong or worth." All of the words have something to do with strength or worth.

Now, knowing that the cluster **val** comes from the Greek word *valere,* try to guess the meanings of the words *before* looking at the definitions in the next section.

covalent _____

_____

invalid _____

_____

valiant _____

_____

equivalent _____

_____

valedictorian _____

_____

validate _____

_____

2.3 Definitions

covalent   in chemistry, covalent refers to the strong bond that is formed between atoms when electrons are shared. (*co* means *with*)

valiant   1. strong, courageous, or brave. 2. performed with heroic achievement.

invalid   1. [in´-və-ləd] weak, sickly, or infirm. 2. [in-va´-ləd] having no force, weight, or strength. (*in* means *not* or *without*)

equivalent   equal in strength or worth. (*equi*, from *aequus*, means *equal*)

valedictorian   the strongest speaker. During a graduation ceremony from high school or college, the valedictorian is the highest ranking student, and he or she delivers a farewell speech during the commencement ceremony. (*dict*, from *dicere*, means *to speak*)

validate   1. to prove or confirm. 2. to make strong by binding under the law.

2.4 Mix and match

Draw lines to connect the words with their meanings.

covalent   strong, courageous, or brave

invalid   equal in strength or worth

valiant   the strongest speaker - usually delivers the commencement speech during the graduation ceremony

equivalent   the strong bond that is formed between atoms when electrons are shared

valedictorian   to prove or confirm

validate   weak, sickly, not strong

2.5 Test yourself

Write the meanings next to the words below.

covalent _____

_____

invalid _____

_____

valiant _____

_____

equivalent _____

_____

valedictorian _____

_____

validate _____

_____

Extra Words

Can you guess the meanings of the following words?

monovalent (*mono* means *one*)

_____

prevalent (*pre* means *before*)

_____

2.6 Using new words

Write sentences using each new word you have learned. If you wish, you may use the words to write a story.

3 Combine Language

3.1 Find the root

Look at the following words:

> combine
>
> compass
>
> decompose
>
> comprehend
>
> comfort
>
> compact

There is a "cluster," or group of three letters, that is exactly the same in all six words. Can you find the cluster?

Circle the cluster that is the same in each word. Write the three letters that make up the cluster. _____ _____ _____

Now look at the words carefully. Because they all have a common cluster, they all have some similarities in their meanings.

Can you guess the meaning of the cluster?

_____

Can you write a definition for any of the words on the list?

_____

_____

_____

_____

3.2 Learn the root

Word Cluster

combine   **com**pass

de**com**pose   **com**prehend

comfort   **com**pact

All of the words above have a common cluster: **com**. The cluster **com** comes from the Latin prefix *com*, which means "with or together." All of the words have something to do with being together.

Now, knowing that the cluster **com** comes from the Latin prefix *com*, try to guess the meanings of the words *before* looking at the definitions in the next section.

combine
_____
_____

compass
_____
_____

decompose
_____
_____

comprehend
_____
_____

comfort
_____
_____

compact
_____
_____

3.3 Definitions

combine   to come together, to make a union, to unite or join. (*bi* means *two*)

compass   1. an instrument for drawing circles. 2. any of various instruments for showing direction. (*passus* means *pace*)

decompose   to break apart. (*de* means *off* or *away*) (*ponere* means *to put* or *place*)

comprehend   to understand, include, take into the mind, or grasp. (*prehendere* means *to seize*)

comfort   1. to strengthen, invigorate, cheer, or enliven. 2. to soothe, relieve, or assist. (*fortis* means *strong*)

compact   closely or firmly united. Compact literally means fastened together. (*pangere* means *to fasten*)

3.4 Mix and match

Draw lines to connect the words with their meanings.

combine   to understand, include, or grasp

compass   closely or firmly united

decompose   to unite together

comprehend   to break apart

comfort   an instrument for showing direction

compact   to strengthen, invigorate, relieve

3.5 Test yourself

Write the meanings next to the words below.

combine _____

_____

compass _____

_____

decompose _____

_____

comprehend _____

_____

comfort _____

_____

compact _____

_____

Extra Words

Can you guess the meanings of the following words?

companion (*pan* means *bread*)

_____

compassion (*pati* means *feel*)

_____

3.6 Using new words

Write sentences using each new word you have learned. If you wish, you may use the words to write a story.

_____

_____

_____

_____

_____

_____

_____

_____

_____

_____

_____

_____

_____

4 Acid Language

4.1 Find the root

Look at the following words:

acid

exacerbate

acrimonious

acute

acerbic

acetometer

There is a "cluster," or group of two letters, that is exactly the same in all six words. Can you find the cluster?

Circle the cluster that is the same in each word. Write the two letters that make up the cluster. _____ _____

Now look at the words carefully. Because they all have a common cluster, they all have some similarities in their meanings.

Can you guess the meaning of the cluster?

_____

Can you write a definition for any of the words on the list?

_____

_____

_____

_____

4.2 Learn the root

Word Cluster

acid   **ac**ute

ex**ac**erbate   **ac**erbic

acrimonious   **ac**etometer

All of the words above have a common word "root": **ac**. The word root **ac** comes from the Latin word *acerbus*, which means "bitter or sharp." All of the words have something to do with being bitter or sharp.

Now, knowing that the cluster **ac** comes from the Latin word *acerbus*, try to guess the meanings of the words *before* looking at the definitions in the next section.

acid _____

_____

exacerbate _____

_____

acrimonious _____

_____

acute _____

_____

acerbic _____

_____

acetometer _____

_____

4.3 Definitions

acid   in chemistry, a substance that is bitter in taste
                    and reacts with a base.

exacerbate   to make more severe or bitter. Exacerbate
                    literally means to make intensively bitter. (*ex*
                    means *intensive*)

acrimonious   severe or sarcastic, especially in speech or
                    temper.

acute   1. ending in a sharp point. 2. keen or quick of
                    mind. 3. severe or sharp pain. 4. an angle under
                    90 degrees.

acerbic   sour, harsh, or severe.

acetometer   an instrument used to measure the amount of
                    acid in a liquid. (*meter* means *to measure*)

4.4 Mix and match

Draw lines to connect the words with their meanings.

acid   to make more severe or bitter

exacerbate   ending in a sharp point

acrimonious   a substance that is bitter in taste and
                        reacts with a base

acute   severe or sarcastic, especially in
                        speech or temper

acerbic   an instrument for measuring the
                        amount of acid in a liquid

acetometer   sour, harsh, or severe

4.5 Test yourself

Write the meanings next to the words below.

acid
_____
_____

exacerbate
_____
_____

acrimonious
_____
_____

acute
_____
_____

acerbic
_____
_____

acetometer
_____
_____

Extra Words

Can you guess the meanings of the following words?

acupuncture (*pungere* means *to pierce*)
_____

acidophilic (*phile* means *to love*)
_____

4.6 Using new words

Write sentences using each new word you have learned. If you wish, you may use the words to write a story.

_____

_____

_____

_____

_____

_____

_____

_____

_____

_____

_____

_____

5 Concentrate Language

5.1 Find the root

Look at the following words:

> concentrate
>
> central
>
> eccentric
>
> egocentric
>
> centrifugal
>
> epicenter

There is a "cluster," or group of four letters, that is exactly the same in all six words. Can you find the cluster?

Circle the cluster that is the same in each word. Write the four letters that make up the cluster. _____ _____ _____ _____

Now look at the words carefully. Because they all have a common cluster, they all have some similarities in their meanings.

Can you guess the meaning of the cluster?

_____

Can you write a definition for any of the words on the list?

_____

_____

_____

_____

5.2 Learn the root

Word Cluster

concentrate   central

eccentric   egocentric

epicenter   centrifugal

All of the words above have a common word "root": **cent**. The word root **cent** comes from the Greek word *kentron*, which means "point or center." All of the words have something to do with being at the point or center.

Now, knowing that the cluster **cent** comes from the Greek word *kentron*, try to guess the meanings of the words *before* looking at the definitions in the next section.

concentrate
_____
_____

eccentric
_____
_____

central
_____
_____

egocentric
_____
_____

epicenter
_____
_____

centrifugal
_____
_____

5.3 Definitions

concentrate
: 1.to collect or focus. 2.to bring to a common center. (*con* means *with* or *together*)

eccentric
: 1. not having a common center. An eccentric circle is a circle that does not have the same center as another circle around it. 2. acting out of the ordinary. 3. deviating from the center. (*ec* means *out*)

central
: 1. in or near the center. 2. main, principle, chief, or basic. 3. equally distant.

egocentric
: self-centered, selfish. An egocentric person is a person preoccupied with himself or herself. (*ego* means *self*)

epicenter
: a focal point, *e.g.,* the central point of an earthquake. (*epi* means *beside, upon, among*)

centrifugal
: fleeing from the center. In physics, it is the force that pushes away from the center during a rotation. (*fugere* means *to flee*)

5.4 Mix and match

Draw lines to connect the words with their meanings.

concentrate   the central point of an earthquake

central   fleeing from the center

eccentric   in or near the center

egocentric   not having the same center

epicenter   self-centered, selfish

centrifugal   to collect or focus

5.5 Test yourself

Write the meanings next to the words below.

concentrate _____

_____

central _____

_____

eccentric _____

_____

egocentric _____

_____

epicenter _____

_____

centrifugal _____

_____

Extra Words

Can you guess the meanings of the following words?

geocentric (*geo* means *earth*) _____

heliocentric (*helio* means *sun*) _____

5.6 Using new words

Write sentences using each new word you have learned. If you wish, you may use the words to write a story.

_____

_____

_____

_____

_____

_____

_____

_____

_____

_____

_____

_____

_____

_____

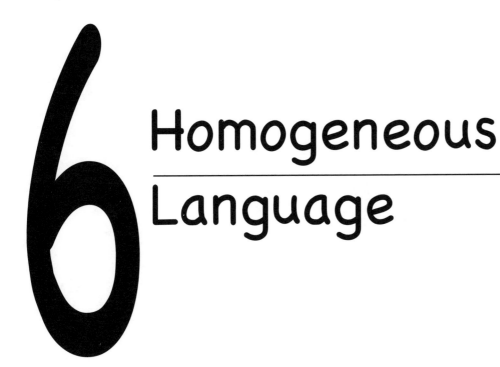

6 Homogeneous Language

6.1 Find the root

Look at the following words:

> homogeneous
>
> general
>
> ingenious
>
> genealogy
>
> hydrogen
>
> pathogen

There is a "cluster," or group of three letters, that is exactly the same in all six words. Can you find the cluster?

Circle the cluster that is the same in each word. Write the three letters that make up the cluster. _____ _____ _____

Now look at the words carefully. Because they all have a common cluster, they all have some similarities in their meanings.

Can you guess the meaning of the cluster?

_____

Can you write a definition for any of the words on the list?

_____

_____

_____

_____

6.2 Learn the root

Word Cluster

homo**gen**eous   **gen**eral

in**gen**ious   **gen**ealogy

hydro**gen**   patho**gen**

All of the words above have a common word "root": **gen**. The word root **gen** comes from the Greek word *genos*, which means "kind, birth, or producing." All of the words have something to do with kind, birth or producing.

Now, knowing that the cluster **gen** comes from the Greek word *genos*, try to guess the meanings of the words *before* looking at the definitions in the next section.

homogeneous _____

_____

general _____

_____

ingenious _____

_____

genealogy _____

_____

hydrogen _____

_____

pathogen _____

_____

6.3 Definitions

homogeneous   of the same kind. (*homo* means *same*)

general   common, ordinary, familiar.

ingenious   1. showing genius, cleverness, and resourcefulness. 2. original and inventive. (*in* means *within*)

genealogy   1. an account or history of the decent of a person. 2. tracking the origin of one's kind. (*logy* means *study of*)

hydrogen   the lightest of all known substances. (*hydro* means *water*)

pathogen   any microscopic organism or virus that can cause disease. (*pathein* means *to suffer*)

6.4 Mix and match

Draw lines to connect the words with their meanings.

homogeneous   the lightest of all known substances

general   an account or history of one's origin

ingenious   of the same kind

genealogy   a microscopic organism that causes
                               diseases

hydrogen   common or ordinary

pathogen   clever, resourceful, original

6.5 Test yourself

Write the meanings next to the words below.

homogeneous _____

_____

general _____

_____

ingenious _____

_____

genealogy _____

_____

hydrogen _____

_____

pathogen _____

_____

Extra Words

Can you guess the meanings of the following words?

nitrogen (*nitro,* from *niter,* means *native soda*)

_____

photogenic (*photo* means *light*)

_____

6.6 Using new words

Write sentences using each new word you have learned. If you wish, you may use the words to write a story.

_____

_____

_____

_____

_____

_____

_____

_____

_____

_____

_____

_____

_____

7 Chromatography Language

7.1 Find the root

Look at the following words:

> chromatography
>
> paragraph
>
> graphology
>
> graphic
>
> photograph
>
> graphite

There is a "cluster," or group of five letters, that is exactly the same in all six words. Can you find the cluster?

Circle the cluster that is the same in each word. Write the five letters that make up the cluster. ___ ___ ___ ___ ___

Now look at the words carefully. Because they all have a common cluster, they all have some similarities in their meanings.

Can you guess the meaning of the cluster?

_____

Can you write a definition for any of the words on the list?

_____

_____

_____

_____

7.2 Learn the root

Word Cluster

chroma**tography**   **graph**ic

graphology   photo**graph**

para**graph**   **graph**ite

All of the words above have a common word "root": **graph**. The word root **graph** comes from the Greek word *graphein*, which means "to write." All of the words have something to do with writing.

Now, knowing that the cluster **graph** comes from the Greek word *graphein*, try to guess the meanings of the words *before* looking at the definitions in the next section.   Look for other word roots you may already know.

chromatography
_____

graphic
_____

graphology
_____

photograph
_____

paragraph
_____

graphite
_____

7.3 Definitions

chromatography   the procedure used to separate different colors.
(*chroma* means *color*)

graphic   1. described in vivid detail. 2. used or expressed
in handwriting. 3. shown by a diagram or plot.

graphology   the study of handwriting to understand a writer's
character. (*logy* means *study of*)

photograph   a picture or image produced by using a
photosensitive surface. (*photo* means *light*)

paragraph   a distinct section or subdivision of a chapter
beginning on a new line and separated by
indentions or spaces from other chapter sections.
(*para* means *alongside* or *resembling*)

graphite   one of the forms of carbon with an iron-gray
color and metallic luster. Graphite is used in the
manufacture of pencils.

7.4 Mix and match

Draw lines to connect the words with their meanings.

chromatography   an iron-gray form of carbon used in
                                   pencils

graphic   described in vivid detail

graphology   technique used to separate color

photograph   a section or subdivision of a chapter

paragraph   a picture recorded using light-sensitive
                                   material

graphite   the study of handwriting

7.5 Test yourself

Write the meanings next to the words below.

chromatography
_____
_____

graphic
_____
_____

graphology
_____
_____

photograph
_____
_____

paragraph
_____
_____

graphite
_____
_____

<u>Extra Words</u>

Can you guess the meanings of the following words?

biography (*bio* means *life*)

_____

telegraphy (*tele* means *from afar* or *distant*)

_____

7.6 Using new words

Write sentences using each new word you have learned. If you wish, you may use the words to write a story.

_____

_____

_____

_____

_____

_____

_____

_____

_____

_____

_____

_____

_____

Carbohydrate Language

8.1 Find the root

Look at the following words:

carbohydrate

dehydrate

hydraulics

hydrochloric

anhydride

hydra

There is a "cluster," or group of four letters, that is exactly the same in all six words. Can you find the cluster?

Circle the cluster that is the same in each word. Write the four letters that make up the cluster. _____ _____ _____ _____

Now look at the words carefully. Because they all have a common cluster, they all have some similarities in their meanings.

Can you guess the meaning of the cluster?

_____

Can you write a definition for any of the words on the list?

_____

_____

_____

8.2 Learn the root

Word Cluster

carbohydrate   dehydrate

hydraulics   hydrochloric

anhydride   hydra

All of the words above have a common word "root": **hydr**. The word root **hydr** comes from the Greek word *hydor*, which means "water." All of the words have something to do with water.

Now, knowing that the cluster **hydr** comes from the Greek word *hydor*, try to guess the meanings of the words *before* looking at the definitions in the next section. Look for other word roots you may already know.

carbohydrate _____

_____

dehydrate _____

_____

hydraulics _____

_____

hydrochloric _____

_____

anhydride _____

_____

hydra _____

_____

8.3 Definitions

carbohydrate | any of several molecules that are made of both water and carbon. (*carbo* means *carbon*)

dehydrate | to remove water from, to dry. (*de* means *away from*)

hydraulics | 1. operated by the force of liquid, *e.g., hydraulic elevator, hydraulic press, hydraulic valve.* 2. setting or hardening under water, as in *hydraulic mortar, hydraulic cement, hydraulic lime.* (*aulos* means *a pipe*)

hydrochloric | designating the acid produced by the combination of hydrogen and chlorine, *i.e., hydrochloric acid.* (*chloro* means *pale green*)

anhydride | any chemical product produced by the removal of water, *e.g., boron oxide is an anhydride.* (*an* means *without*)

hydra | 1. a small fresh-water animal with a tube-like body. 2. in Greek mythology, the nine-headed serpent that was slain by Hercules.

8.4 Mix and match

Draw lines to connect the words with their meanings.

carbohydrate   a small fresh water animal with a tube-like body

dehydrate   operated by the force of liquid

hydraulics   a molecule made of both water and carbon

hydrochloric   to remove water from, to dry

anhydride   designating the acid produced by hydrogen and chlorine

hydra   a chemical product produced by the removal of water

8.5 Test yourself

Write the meanings next to the words below.

carbohydrate
_____

_____

dehydrate
_____

_____

hydraulics
_____

_____

hydrochloric
_____

_____

anhydride
_____

_____

hydra
_____

_____

Extra Words

Can you guess the meanings of the following words?

hydroplane (*plane* means *level surface*)

_____

hydrophobia (*phobia* means *fear*)

_____

8.6 Using new words

Write sentences using each new word you have learned. If you wish, you may use the words to write a story.

_____

_____

_____

_____

_____

_____

_____

_____

_____

_____

_____

_____

_____

_____

9 Polymer Language

9.1 Find the root

Look at the following words:

> polymer
>
> polyester
>
> polygon
>
> polygraph
>
> polynomial
>
> Polynesia

There is a "cluster," or group of four letters, that is exactly the same in all six words. Can you find the cluster?

Circle the cluster that is the same in each word. Write the four letters that make up the cluster. _____ _____ _____ _____

Now look at the words carefully. Because they all have a common cluster, they all have some similarities in their meanings.

Can you guess the meaning of the cluster?

_____

Can you write a definition for any of the words on the list?

_____

_____

_____

9.2 Learn the root

Word Cluster

polymer	**poly**ester
polygon	**poly**graph
polynomial	**Poly**nesia

All of the words above have a common word "root": **poly**. The word root **poly** comes from the Greek word *polys*, which means "many." All of the words have something to do with many.

Now, knowing that the cluster **poly** comes from the Greek word *polys,* try to guess the meanings of the words *before* looking at the definitions in the next section. Look for other word roots you may already know.

polymer
_____

_____

polyester
_____

_____

polygon
_____

_____

polynomial
_____

_____

polygraph
_____

_____

Polynesia
_____

_____

9.3 Definitions

polymer

in chemistry, a compound made of repeating units. (*mer* means *unit*)

polyester

a specific type of polymer made by condensing polyhydric alcohols. (esters)

polygon

in geometry, a plane figure with many angles and sides. (*gonia* means *angle*)

polygraph

1. a device for reproducing writings or drawings.
2. a person who writes many works or many kinds of works. (*graphein* means *to write*)

polynomial

1. containing many names or terms. 2. in algebra, an equation with many terms, *e.g.*, $x^3 + x^2 + x + 2 = y$. (*nomos* means *law*)

Polynesia

a great number of islands in the Pacific Ocean, east of the international date line, extending from New Zealand in the south to the Hawaiian Islands in the north.

9.4 Mix and match

Draw lines to connect the words with their meanings.

polymer   a group of islands in the Pacific Ocean

polyester   a plane figure with many sides and
                           angles

polygon   in chemistry, a compound composed of
                           many repeating units

polygraph   containing many names or terms

polynomial   a specific polymer made of many
                           esters

Polynesia   a device used for reproducing writings

9.5 Test yourself

Write the meanings next to the words below.

polymer
_____
_____

polyester
_____
_____

polygon
_____
_____

polygraph
_____
_____

polynomial
_____
_____

Polynesia
_____
_____

<u>Extra Words</u>

Can you guess the meanings of the following words?

polysaccharide (*saccharide* means *sugar*)
_____

polytheism (*theos* means *god*)
_____

9.6 Using new words

Write sentences using each new word you have learned. If you wish, you may use the words to write a story.

_____

_____

_____

_____

_____

_____

_____

_____

_____

_____

_____

_____

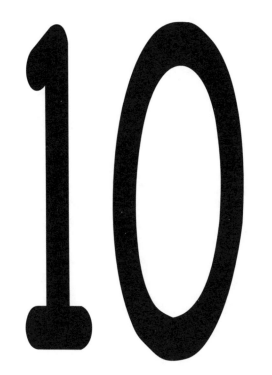

Language
Protein

10.1 Find the root

Look at the following words:

> protein
>
> protocol
>
> prototype
>
> protozoa
>
> proton
>
> protagonist

There is a "cluster," or group of four letters, that is exactly the same in all six words. Can you find the cluster?

Circle the cluster that is the same in each word. Write the four letters that make up the cluster. _____ _____ _____ _____

Now look at the words carefully. Because they all have a common cluster, they all have some similarities in their meanings.

Can you guess the meaning of the cluster?

_____

Can you write a definition for any of the words on the list?

_____

_____

_____

_____

10.2 Learn the root

Word Cluster

protein   **prot**ocol

prototype   **prot**ozoa

proton   **prot**agonist

All of the words above have a common word "root": **prot**. The word root **prot** comes from the Greek word *protos*, which means "first, early." All of the words have something to do with being first or early.

Now, knowing that the cluster **prot** comes from the Greek word *protos,* try to guess the meanings of the words *before* looking at the definitions in the next section.   Look for other word roots you may already know.

protein
_____
_____

prototype
_____
_____

protocol
_____
_____

protozoa
_____
_____

proton
_____
_____

protagonist
_____
_____

10.3 Definitions

protein	a polymer chain of repeating amino acids. It literally means "to come first" or "first rank."
protocol	the first sheet of a legal instrument which was glued to the document. (*kolla* means *glue*)
prototype	an original or first model, the first of its kind. (*typos* means *form* or *model*)
protozoa	microscopic one-celled organisms belonging to the kingdom Protista. Protozoa can have both animal-like and plant-like characteristics, but were considered "first animals." (*zo* means *animal*)
proton	fundamental particle of an atom with atomic weight of 1 amu and found in the nucleus.
protagonist	the primary (first) or leading character or actor in a play, novel, or story. (*agonistes* means *actor*)

10.4 Mix and match

Draw lines to connect the words with their meanings.

protein   microscopic one-celled organisms

protocol   fundamental particle of an atom

prototype   a polymer chain of repeating amino acids

protozoa   the primary character in a play

proton   the first sheet glued to a document

protagonist   the original model of its kind

10.5 Test yourself

Write the meanings next to the words below.

protein
_____
_____

protocol
_____

prototype
_____

protozoa
_____

proton
_____

protagonist
_____
_____

Extra Words

Can you guess the meanings of the following words?

protolithic (_lithos_ means _stone_)

_____

prototrophic (_troph_ means _nourish_)

_____

10.6   Using new words

Write sentences using each new word you have learned. If you wish, you may use the words to write a story.

_____

_____

_____

_____

_____

_____

_____

_____

_____

_____

_____

_____

_____

_____